The AI Survival Guide

Navigating Displacements in IT and Other Industries and Succeeding in a New World of Artificial Intelligence

Andy Maslov, PhD, MBA

CelebrityTechie.com

CONTENTS

Dedicated to celebration of life

About the Author and This Book

My life has undergone profound personal and professional transformations, offering a journey brimming with unconventional dynamics and rollercoasters. I plan to explore these experiences further in my upcoming writings.

From an early age, I have embraced a life filled with adventure, traveling and working across multiple continents. My enthusiasm for technology and business has remained steadfast throughout this semi-nomadic lifestyle.

Driven by a desire to learn more about and excel in business, I pursued and achieved an MBA from the Kellogg School of Management in Illinois. This endeavor greatly enhanced my comprehension of the business world and reinforced my perspective on technology as a crucial business enabler and driver.

My engagement with technology and business has evolved from programming to ERP consulting, strategic business advisory, project management, product development, technology architecture, executive leadership, entrepreneurship, serving on boards of directors, and investing in ventures.

This journey, among other things, has endowed me with a unique and comprehensive understanding of technology, encompassing software development, architecture, cloud computing, Machine Learning (ML) and Artificial Intelligence (AI), Distributed Ledger Technology (DLT) and crypto, cybersecurity, business applications, high-performance and mission-critical systems, interactive technologies, and technology strategy, among many other areas.

One of the most noteworthy accomplishments in my career is the development of a uniquely effective system of methods for consulting and coaching business leaders on Enterprise Information Technologies, enabling them to achieve swift results with unparalleled financial leverage.

At CelebrityTechie.com, we use and continuously refine this system to offer exclusive consultancy and coaching to top business leaders on technological matters. We have also established an Academy designed to guide aspiring students toward becoming elite-level celebrity professionals.

We pride ourselves on our ability to address challenges of any complexity, offering top-level advisory, coaching, and training that immensely enhance businesses, careers, and personal growth. With our distinctive expertise, we can quickly turn you into a technologically superior business shark!

Our system is especially crucial in this new era of AI proliferation, where elite professionals equipped with AI and other automation tools will dominate entire industries. It is most effectively taught through visual and interactive engagement with real-life business contexts and structured consideration of a wide range of technology aspects and options.

While this book does not focus on training using our system, its primary aim is to serve as an alert, highlighting AI's profound and growing impact on our lives, outlining the available options, and emphasizing the urgent need to develop a new mindset to navigate these rapid changes.

INTRODUCTION

This book spotlights the technology field, otherwise known as Information Technologies (IT) or Enterprise Information Technologies. However, its relevance extends to most professions, businesses, and industries. Regardless of your career path, you will gain valuable knowledge from it.

In today's world, we are all techies. It is simply impossible to escape the influence of technologies in our daily lives, regardless of where we go or what we do. The ability to skillfully handle and navigate technologies has become an essential proficiency in numerous occupations.

In fact, possessing superior technology expertise provides a distinct advantage and monetizes handsomely in our professional lives. Technical knowledge and skills are essential assets when starting a new venture as an entrepreneur.

Elon Musk even recommends studying engineering as one of the few surviving professions in the near future (Gilchrist, 2019).

As written by a seasoned practitioner with firsthand experience, this book goes beyond bland or overly optimistic narratives. It delves into current and anticipated developments, offering practical guidance for the immediate future based on real-world insights.

The impending Artificial Intelligence era promises to simultaneously be marvelous and daunting for humanity, offering a blend of outstanding opportunities and unprecedented challenges.

The rapid advancement of technologies has been causing significant disruptions in the job market, resulting in widespread workforce displacement at all levels. Around the world, policies intended to address and mitigate these changes need more help to keep up.

In this rapidly changing world, two key possibilities lie ahead for individuals, regardless of how public policies may evolve to mitigate the impact and help society adapt.

The first possibility is striving to achieve the premier league level in the profession. This becomes increasingly urgent for knowledge-based occupations such as IT, law, finance, healthcare, and others. More than simply acquiring essential knowledge and skills is required. Professionals must not only master AI and other automation tools and services but also indefinitely advance their core expertise to remain relevant and effectively use new automation instruments.

The second possibility is to explore other opportunities, such as different industries, careers, and entrepreneurial ventures. This is particularly relevant for those who may feel that their current profession is at a higher risk of being taken over by automation. Exploring alternative opportunities can help individuals stay ahead of the curve and remain in demand in the changing economy. This in itself is a skill that needs to be mastered.

It is a tough choice, and both possibilities are challenging to pursue.

The hard reality is that numerous knowledge-based and office jobs, including IT, will require fewer people now. Exceptional qualifications, brightness, and creativity are necessary for those who stay ahead, along with a willingness to expand their horizons and think and act cross-functionally.

Many people have considered IT a safer harbor and more immune to AI and automation takeover than other areas and industries. For many, it has become a highly sought-after ticket to becoming a citizen of the world, having a stable income, and possibly more. Some are turning to IT from other professions, trying to secure their livelihood for the foreseeable future, where technologies will continue to play a growing role as enablers for many other industries and different aspects of our lives.

Although this view regarding the technology industry was once essentially correct, it is now quite demanding to work in IT, staying up to date with all of the fast-paced advancements and increasing complexity, and it will keep getting even more challenging. The incumbent engineers often complain about chronic fatigue and burnout nowadays, which is a real problem (Cook, 2015; White, 2023).

And now, with the coming of AI, to be a sought-after techie, you will need to become one of the best. As Andrew Yang notices in his book *The War on Normal People*, "... in Manhattan and Silicon Valley and Washington, DC, my friends and I will be busier than ever fighting to stay current and climb within our own hypercompetitive environments" (Yang, 2019).

The army of outsourced "commodity" engineers is quickly becoming a thing of the past. The industry is undergoing its own immense transformation.

The same wind is picking up in all professions and industries, threatening to become a hurricane. Hurricanes in the Atlantic and Pacific often get names, and we can call this one Hurricane AI.

The coming era is the era of premier-level professionals, requiring a more thoughtful choice of occupation and focused preparation for it. Newbies and industry incumbents must become different, with different mindsets, attitudes, training, and education. They will face the tough choice of doing it fast or considering alternatives.

AI becomes an increasingly more valuable and integral tool in many professions, but it also raises the bar and sets higher expectations for those who use it.

This will profoundly change the workforce and professions, and the book shows how to prepare, use the benefits, and overcome the challenges of the coming era for engineers, consultants, analysts, lawyers, bankers, healthcare professionals, managers, recruiters, entrepreneurs, investors, business owners, executives, and board directors—essentially for everyone.

To stay in business (and better yet, to thrive), you must become a premier leaguer in your occupation and do it fast.

Premier-level professionals equipped with AI can do more, better, and faster. Not being knowledgeable and skilled enough in a profession can cause more damage than good, as AI can immensely amplify errors.

AI cannot replace the need for human expertise. In fact, AI technology requires a higher level of expertise from its users to ensure that the output of AI is thoroughly assessed, relevant, and effective. Therefore, it is essential for individuals to have a deep and comprehensive understanding of their field of work in order to utilize AI technology effectively.

With the coming of AI, becoming a professional, especially in knowledge-based occupations, will get increasingly similar to becoming a professional athlete.

Although I have been involved in technology and business since childhood, building and selling computers and programming in high school, my personal journey in the field has not been easy and involved a lot of hard work and learning through exploration and experimentation.

I have pursued a diverse education and training in computer science and business, including obtaining an MBA from a top business school and a Ph.D. in Business Economics.

Having been practicing from my early years at the university, I have tried it all in the trenches, worked in multiple countries, primarily Canada, the US, and Australia, with numerous different teams, explored all possible venues in the technology industry (and there are myriads of them), worked in Big4 consultancies, technology giants, with biggest banks, energy, mining, retail, manufacturing, and telecom companies and in the highly agile and entrepreneurial fintech sector, created my own startup companies, worked as a technology executive and invested in ventures.

Becoming a premier-level professional, techie, or business leader is an arduous journey that requires unwavering determination and a strong will to succeed. In today's world, this path is not just desirable but essential to achieve success. It is imperative to embrace this challenge and make it a top priority.

Despite the challenging times, this book is charged with optimism, though. Even with AI's proliferation and increasing capabilities, the human mind will always have a unique potential to flourish, invent, create, and make a difference. It depends on us how we forge and further develop this ability moving forward.

I could spend an infinite amount of time writing this book, which is an ongoing story, continuously update it with new facts and insights, and connect more dots. However, eventually, you must draw a line and finish the writing when it encompasses your key thoughts on the subject.

I made the book compact on purpose. The most powerful messages are simple and concise, and I do not want to overwhelm you, the reader, with too much anecdotal information to get the messages across. At the same time, it is well-researched, and I supported my writing with necessary references where appropriate. This is a brief but one of the most relevant books that you would read, no matter what you currently do.

This book represents my personal opinions and perspectives. The views expressed herein are solely my own and do not necessarily reflect those of any other individual or organization.

I have been fortunate to work with exceptional executive leaders and business owners who have driven numerous remarkable achievements in the business and technology sectors. Their inspirational guidance has been truly impactful, and I am grateful for this ongoing experience.

If you are interested in exploring how to capitalize on exciting new opportunities and overcome challenges in a world rapidly reshaped by AI and automation, please do not hesitate to contact us at www.CelebrityTechie.com.

References

Gilchrist, K. (2019, August 29). Elon Musk: AI will make jobs 'kind of pointless' — so study this. CNBC.

Cook, S. (2015). Job burnout of information technology workers. International Journal of Business, Humanities and Technology, 5(3).

White, S. (2023, November 3). Burnout: An IT epidemic in the making. CIO.

Yang, A. (2019). The war on normal people: The truth about America's disappearing jobs and why universal basic income is our future. Hachette Books.

At the new AI frontier

In the not-so-distant past, recruiters relentlessly looked for "good candidates" in IT professions on desperate requests from businesses. However, lately, a different trend has been noticed, where hiring managers are bombarded by many recruiters and all sorts of outsourcing providers offering technology people for hire.

While not surprising, this is very symptomatic and is just one of the early signs of future developments. Some might attribute this to the consequences of the COVID-19 times and the ongoing turbulent economic conditions worldwide. But there are also more transformational forces in play.

Technology giants are laying off in troves while just recently luring engineers fiercely from each other, deploying huge and fancy amenities, and adding all sorts of entertainment and perks for the staff. Mark Zuckerberg defined 2023 as "the year of efficiency" and said that the company's round of 11,000 layoffs was only the tip of the iceberg, with more to come (Weinberger, 2023). More tech companies just recently announced new waves of layoffs after more than 260,000 workers lost their jobs across the industry last year (Chadwick, 2024).

Locations traditionally known for outsourcing are anticipated to be increasingly impacted (Verma, 2023).

Many technologies and solutions have already been created and just need maintenance and enhancement. The day-to-day job of most techies now is to fix bugs and other issues in existing solutions and add extra features, which is called "brownfield" work.

The term is borrowed from urban planning, where "brownfield" sites are previously developed land that may be contaminated or have existing structures, as opposed to "greenfield" sites, which are undeveloped and free from constraints.

Not so many are lucky enough to participate in creating something new in the "greenfield." And then, not so many turn out to be prepared and capable of producing quality new systems in the "greenfield," using the benefits of the latest advancements and developments made available in the technology industry.

The industry has become overwhelmed with participants with patched and fragmented expertise, poor levels of knowledge, and sometimes even bluntly con artists. This happens partly because, with the ever-increasing complexity of the technology landscape, it becomes increasingly more difficult for businesses (including technology ones) to assess techies. As a result, businesses and organizations get websites that work poorly, delays, failures, and errors in processing data, systems being easily hacked, etc., with rapidly growing technology costs at the same time.

Quality has been a big elephant in the room and an immense pain in the technology industry and well beyond. There was a golden brief period when the work in IT was considered artful and even magical in a sense. Then, things deteriorated fast with the rise of cheap outsourcing and related commoditization and devaluation of the technology workforce. In pursuit of lower costs, businesses have actually spent much more money in the long term, considering chronic delays in delivery, costs of rework, and overblown technical debt dragging them down as a huge rusty anchor.

With the devaluation of the technology workforce, both the quality of the workforce and the quality of the products produced in IT have been dropping. These issues affect the ongoing delivery of products and services and even slow down, if not block, the overall societal progress.

Some industry voices' appeal for craftsmanship in the technology profession largely went unheeded. The snowball of technical debt across all industries and businesses keeps growing.

The technical debt stems from creating poorly planned, architected, poorly developed, and implemented solutions. It causes increased expenses required to support such solutions, integrate them, and simply use them immediately and in the long term.

Too often, a freshly implemented solution is considered scrap almost immediately and requires replacement. Similarly, projects that span multiple years, if not decades, may fail to launch into production or only achieve partial success.

This has a broader effect on the economy. While many people were able to earn a living in the technology industry, consumers at large suffered from increased costs and prices due to the vast technical debt accumulated in the economy.

According to the Global Technology Executive Survey (Protiviti, n.d.), "...nearly 70% of organizations view technical debt as having a high level of impact on their ability to innovate."

Peter Thiel (2014) cited a lack of true innovation in recent decades. Even AI cannot be considered much of an innovation, as it primarily consists of known mathematical and algorithmic methods that are now made possible to process large quantities of data on more performant technology infrastructure.

For example, artificial neural networks, particularly deep learning, have been at the forefront of AI's recent advancements in fields such as image and speech recognition, natural language processing, autonomous vehicles, etc. They are called neural networks because they are inspired by the structure and function of biological neural networks in the brain.

At the same time, the history of artificial neural networks dates back to the math of 18-19th century (Wikipedia contributors, n.d.). I developed an artificial neural network for forecasting as my graduation thesis in Computer Science during my university years.

As Thiel elaborates further in his book, various reasons have contributed to the lack of genuine innovations over the past several decades. However, the burden of technical debt is a major factor as well.

After Elon Musk took over Twitter, he quickly figured out the issue with the platform's technical debt. Observing system failures following introduced changes, he concluded that the code was brittle and required a rewrite (Griffin, 2023). Some attributed these issues to recent layoffs. However, the fact that the system failed in the absence of certain employees suggests that the architecture and setup were inadequate in preventing such scenarios. You can imagine how significant the issue of technical debt is that even prominent Silicon Valley companies could not avoid its accumulation.

Technical debt is produced not only in custom software development. Even if there is an excellent off-the-shelf commercial software or SaaS product, not qualified enough implementers can make its implementation a complete disaster. This happens all too often, as you can hear, "We failed to implement ERP system A, so now we scrap it and will try another ERP system B, C, D ..."

The seemingly lower barrier to entry into IT was a significant factor. Many thought like this: Okay, what do I need to do to get into the profession? It is simple, right? Just learn some programming language that already resembles the natural language I can speak, and then bingo!

Guaranteed placement and good pay in many cases, and a "work and travel" package for those who enjoy exploring the world or strive to move to the parts of the world considered to have better economies.

When many new systems were being developed in the earlier period of IT history and demand for anyone claiming to be an engineer was high, that was passable.

The IT industry has remained a sanctuary for so long, though, and it now gets overdue for rebalancing. Artificial Intelligence becomes a great catalyst in this process.

The demand for quality remains as high as ever, and being a techie requires being as artful as ever.

However, many industry participants are up for the challenge, which now requires a greater deal of expertise and concentration, like in a professional sport.

In the past, the professions of accountant and lawyer were sure bets for earning money. Then came information technologies. But now AI is crashing that world for everyone, for good or bad.

How would AI affect the technology industry that gave it birth in the first place?

Even at this early stage, AI is already assisting in rewriting and improving legacy applications, adding testing coverage, writing boilerplate code, and many other laborious tasks.

As Vitalik Buterin correctly points out, AI is already able to locate and fix defects (or "bugs") in all kinds of software (Buterin, 2024). Many teams used to specialize in this traditionally laborious task.

As you can gather from these examples, the expanding list of AI applications in IT is effectively reducing the scope of work previously acceptable for engineers with lower qualifications, placing more emphasis now on using premier-level techies equipped with AI.

This will create many redundancies in the IT industry and demand more from IT training beyond what traditional academy or the existing training industry can provide. Due to reduced demand and growing competition, the technology workforce is in for a challenging time. The industry participants are braced for a rough awakening.

What about other industries? Most of the traditional office and knowledge-based occupations you can think of will make up candidates for AI takeover to various extents.

For example, many people are looking forward to AI becoming a cheaper and more affordable means of managing their legal affairs and legal protection in courts. This is an area that is a perfect match for generative AI.

To survive and succeed as lawyers, legal professionals will need to excel in their field, possess the ability to teach, guide, and supervise artificial intelligence for the cases they handle, and focus on becoming skilled in their clients' legal affairs and representation while also being thought leaders in the legal profession.

Even the wealthiest clients who used to hire multiple legal teams while having all the necessary financial means at their disposal have already started using AI legal services to consult them on their legal cases, reducing billed hours spent with external lawyers.

Automating due-diligence review of commercial loan agreements already saves JPMorgan Chase 360,000 hours of work previously spent each year by lawyers and loan officers (Son, 2017).

In his memo to employees, the head of one of Europe's biggest media groups Mathias Döpfner warned journalists that AI could take over their jobs. His further statement aligns precisely with the core theme of this book, which is the ongoing formation and dominance of top-tier professionals in professions during the era of AI: "In short, the creation of exclusive and attractive content remains irreplaceable and is going to become even more critical to success for publishers. ... **Only those who create the best original content will survive.**" (Cooban, 2023)

Even DJs get slashed, with taking advantage of investments in technology and artificial intelligence being cited as the primary reason, as was the case at iHeartMedia (Harwell, 2020).

AI and automation are increasingly facilitating operations and analytics in accounting and auditing. Technology firms are on a quest to capture a larger share of the industry, where the Big Four alone earned USD 190 billion in 2022. Companies in the Crunchbase accounting niche secured nearly a billion U.S. dollars in early investments, highlighting the technology industry's efforts to further automate accounting and auditing professions (Glasner, 2024).

Industry veterans are expressing their concerns about the future of the accounting profession. As an author sharply stated in her article, "The future of accountants is uncertain, and I'm not optimistic. As AI models like ChatGPT continue to evolve, we risk becoming increasingly obsolete. The era of sitting in a cubicle and crunching numbers may soon end. We must adapt or risk being left behind. We can either embrace the machines and learn to work alongside them or watch as our jobs vanish. The choice is ours" (Gonzalez, 2023).

In healthcare, diagnosing human health based on symptoms and test results, including X-ray, ultrasound,

CT scan, and MRI image diagnostics, is well suited for what AI can do effectively.

ChatGPT has already successfully passed the United States Medical Licensing Examination (USMLE), a three-step test series that medical graduates must pass to practice medicine in the U.S. (James, 2023). Although, this presents a potential controversy. When AI takes an exam, it essentially operates in an "open-book" mode, as it has unlimited access to the knowledge it obtained from textbooks, articles, and other materials loaded into it.

Financial consultants – checked off.

And the list goes on and on ...

No one's status quo is safe, including top echelons across businesses. All levels of management are particularly vulnerable as well. At an earlier stage in its growth, Google had already experimented with removing most of the management layers (Garvin, 2013). Their experiment did not go well back then, and they quickly rolled back on that, but with the coming of AI, that can be revisited now.

In this context, executives and board members are increasingly expected to possess higher levels of proficiency in technology.

According to a global executive search firm, having in-depth technology expertise is becoming crucial for CEOs. However, finding such expertise remains challenging, leading to many companies not being adequately up-to-date with AI and technology advancements (Dinte Global Executive Search, n.d.).

This is also true for other business executives and is often relevant to top technologists such as CIOs and CTOs.

Research conducted by the David and Sharon Johnston Centre for Corporate Governance Innovation underscores that "... board composition in Canada and the U.S. must evolve more quickly to ensure that the right team of directors with the appropriate expertise is chosen for the long-term survival of a company. Companies that lack directors with technological expertise are in danger of costly mistakes when they develop their new technological strategies and systems" (Officer, 2022).

AI will be causing redundancies across industries and occupations, and the best will have a higher chance to endure and prosper by producing artful work.

Thomas Dohmke, CEO of GitHub, recently admitted that about half of the code produced by users of Copilot is already AI-generated (Knight, 2024). He also correctly notices that AI still needs to be operated by humans, but the fact of the matter is that fewer people are now required to write software code.

Half of the code already! And this is just one of the rapidly growing number of AI tools. And this is just the beginning. You can do the math for productivity and jobs.

The development of new business ideas should be accelerated to create more opportunities and offset the decrease in demand for engineers caused by the rise of AI-assisted productivity. But that can be expected to be lagging.

This leads to a remarkable workforce redistribution. As discussed in the book, the primary options are: becoming one of the best in a profession or looking for opportunities in different fields and niches.

As often happens with new revolutionary technologies, AI is not universally one-sided on benefits. There are crucial reservations with a broad impact on individuals and society at large.

References

Weinberger, M. (2023, February 3). Mark Zuckerberg just ushered in a new era of tech, where profitability and efficiency trump perks and culture. Business Insider.

Chadwick, L. (2024, January 25). Tech companies are laying off thousands of workers: What does this mean for the industry? Euronews.

Verma, P. (2023, October 30). Indians more afraid of AI taking away jobs than their global peers. The Economic Times.

Protiviti. (n.d.). Global technology executive survey: Tech debt a major burden. Protiviti.

Thiel, P. (2014). Zero to one: Notes on startups, or how to build the future. Crown Currency.

Wikipedia contributors. (n.d.). History of artificial neural networks. Wikipedia.

Griffin, A. (2023, March 8). Twitter needs a 'complete rewrite' after it broke again, Elon Musk says. The Independent.

Buterin, V. [@VitalikButerin]. (2023, March 3). One application of AI that I am excited about is AI-assisted formal verification of code and bug finding... [Tweet]. Twitter.

Son, H. (2017, February 28). JPMorgan software does in seconds what took lawyers 360,000 hours. Bloomberg.

Cooban, A. (2023, March 1). The owner of Insider and Politico tells journalists: AI is coming for your jobs. CNN.

Harwell, D. (2020, January 31). iHeartMedia laid off hundreds of radio DJs. Executives blame AI. DJs blame the executives. The Washington Post.

Glasner, J. (2024, February 1). AI will be doing more accounting if startup investors have their way. Crunchbase News.

Gonzalez, A. (2023, March 29). ChatGPT talks about AI taking accountants' jobs from the perspective of an accountant losing their job to AI. Going Concern.

James, T. A. (2023, April 13). How artificial intelligence is disrupting medicine and what it means for physicians. Harvard Medical School Postgraduate Medical Education.

Garvin, D. A. (2013, December). How Google sold its engineers on management. Harvard Business Review.

Dinte Global Executive Search. (n.d.). Finding the right CEO in the age of technology is no easy task.

Officer, S. (2022, July 26). Why do corporate boards need directors with technological skills? David and Sharon

Johnston Centre for Corporate Governance Innovation, Rotman School of Management, University of Toronto.

Knight, W. (2024, February 8). AI Is Rewiring Coders' Brains. Yours May Be Next. Wired.

AI HITTING LIMITS

Of course, AI is not all mighty. It is far from it and will remain that far for quite some time, and still will not master all the capabilities of humans.

It makes mistakes, gets confused, confuses, mixes old and new information, mixes incompatible things, makes you go in endless circles, and so forth.

Such occasional incoherent behavior of AI tools is called AI hallucinations (Wikipedia contributors, n.d.). You cannot entirely rely on a tool that has occasional unpredictable hallucinations.

It can even spiral into a downright bizarre behavior (Eaton, 2024).

Another distinct feature of generative AI is that it often yields different responses to the same question, owing to the dynamic nature of its models and knowledge base, randomness in the model's response generation process, as well as its attention to the exact wording and varying contexts of the question.

There are a few other critical factors to mention regarding the use of AI.

First, input: What data is being used to train AI? If some knowledge was not fed into AI in its learning stage, that knowledge would not be available for AI to process and use.

Second, access and output: The offering of AI is limited by the parties who own and operate it. Those parties can increase the price for using AI, which can become an access barrier. They can restrict the use rate, filter your input, not allow specific requests, and filter or modify AI output (programmatically, based on algorithms). A more benign example of such limitation is when you try to ask the GitHub Copilot AI tool a non-technical question, but it refuses to answer. I have also encountered an occasional behavior in this tool where when you use certain words, such as "execution," in the context of executing an application, it might refuse to answer, referring to some "policies" and asking for feedback in case it wrongly detected a "policy violation." It can be an odd and puzzling experience.

Third – skill and knowledge lock-in: This will start being felt pretty quickly. People will get increasingly hooked and reliant on AI tools at work and home. Retaining certain skills and abilities that were the norm before will become challenging. It is similar to what happened when we started using calculators, spreadsheets, and payment cards, not requiring us to calculate change money or do other simple calculations in our minds. Most people have just lost that skill to a large degree. This quality of AI is not its limitation per se but rather a detrimental consequence of its utility.

While AI can be good at boilerplating and prototyping solutions in various fields, it may encounter difficulties when making conclusions and suggestions for specific business and solution contexts. There are myriads of such contexts, with a virtually unlimited variety of possible combinations that AI cannot learn exhaustively.

Further comparing capabilities of humans and AI: Humans may have insufficient knowledge and incorrectly or ineffectively use the knowledge they have. However, the same holds true for AI, as it may also have inadequate or incorrect knowledge and can use it incorrectly and ineffectively (e.g., due to its algorithmic nature). In many cases, you will not know what exact knowledge it uses and from what sources. And even if you use your own AI model and train it with your own data, you will not be certain about the exact knowledge derivation mechanics it uses.

Fundamentally, it can be said that AI is missing a "soul." For example, when it comes to book writing, AI will not write a truly meaningful book for you. Your book needs your inspiration, character, and depth, and its writing largely relies on your unique and rich experience and background that are not available for AI to learn from in detail.

AI is not creative. One cannot expect sparks of ingenious creativity, true wonders, or mystical experiences from it (if we do not consider AI hallucinations mystical). It operates with the knowledge it consumes when "learning," and the maximum creativity it can achieve is from mixing and matching that knowledge.

Here comes another pitfall. The quality of AI depends on the quality of information it is fed with when learning.

If certain information is unavailable to AI, there is little chance it can infer that knowledge independently. AI may perform suboptimally because it has not learned certain information, particularly if it is not accessible to people providing training knowledge to the AI.

Information used by AI for learning comes from a wide variety of sources, including websites, books, articles, and other written, graphical, or sound materials. All of these are ultimately products of content creators under various copyright models.

What motivates content creators? Factors such as publicity and income play a vital role, essentially stemming from the association of their personality with the materials they publish.

What does AI do? It effectively anonymizes knowledge. It extracts information from all available sources it is fed and processes that information into synthesized "knowledge" without attributing sources. As such, attribution is challenging to trace through AI algorithms.

Here comes what I call **a content creation dilemma**: People are increasingly using AI and turning less to original content. Without direct attention from human users and attribution, content creators may lose motivation to create new quality content for consumption. As a result, AI receives less actual knowledge to learn from and becomes less knowledgeable and useful. This mechanism may become self-correcting, though. As the quality of AI output decreases, people may turn more to original content, encouraging content creators to produce it.

In this context, search engines remain gateways to originally authored content on the Internet, while they may be used less now, with AI serving a growing part of user inquiries.

Professionals should exercise caution while using AI, as it can make mistakes ranging in severity. For example, incorrect coding advice from AI taken for granted can lead to serious flaws in software and technical debt.

This especially impacts less knowledgeable professionals as it would be difficult for them to recognize the issues in AI output, especially considering that advice from AI often comes in a prescriptive and authoritative form, which can give an illusion that this is always the right answer.

Solutions based on unfiltered advice are often flawed and suboptimal. The rapid pace of creation facilitated by AI assistance further compounds this issue, resulting in a larger volume of work that needs thorough examination and rectification.

At the same time, more knowledgeable professionals will be getting increasingly self-reliant as they can more effectively cover a broader scope with assistance from AI.

Therefore, competition for qualified roles will be heating up with rapidly increasing productivity.

AI has reduced the need for less knowledgeable individuals to assist more knowledgeable professionals in preparatory tasks.

This has already been noticed by professionals in many industries. For example, practitioners in the law industry share their observations on Linkedin and other social media, reporting that generative AI enables partners in law firms to do more now with fewer associates. The same applies to consulting. The entry barriers for graduates and junior professionals are rapidly increasing.

AI is rapidly advancing, overcoming its ongoing limitations and providing powerful tools across various fields. As these constraints diminish, the potential for AI to profoundly transform our lives increases.

References

Wikipedia contributors. (n.d.). Hallucination (artificial intelligence). Wikipedia.

Eaton, K. (2024, February 29). Microsoft's Copilot offers bizarre, bullying responses: Latest AI flaw. Inc.

WHO IS A CELEBRITY PROFESSIONAL?

If someone is committed to becoming a professional or staying and thriving as a professional in the coming age, they must strive to reach the premier league. Figuratively speaking, they have got to eventually become a "celebrity professional" or "celebrity techie" in the context of the technology realm.

Let us consider the making of a celebrity techie, which can be extrapolated to many other occupations affected by AI.

The celebrity techie is a creator and savior for businesses that employ or contract him. They drive startups through early challenges and help mature companies reshape and prosper.

It eventually comes down to passion and talent. Not everyone who enters technology has a special vigor or inclination for it. Many just want to become engineers to reliably earn a living, on a road that has already been well-tried by at least a couple of generations now. And it might have been expected in the past. But that changes now.

Too much work has been done at a mediocre level or worse. We have a possibility of changing this around and raising the quality of life to an entirely different level.

Talent is essentially a product of hard work. They say, "Work smart, not hard." This is not entirely true. To become a celebrity-level professional, you must work hard and smart.

And not just that. There is also another secret ingredient. It is an inner compass and "sense of smell" for what can be considered a piece of art in your profession or business or what stinks. This component is especially hard to master, as can be observed in many spheres. Good software developers are trained to sense when "code smells," as a simplified example. It is not only code that can smell in software engineering. Things may start rotting from the levels of strategy, roadmaps, Project Management Office (PMO), and architecture and proliferate all the way through all facets of the products and services being delivered.

The expertise of a celebrity professional is well-rounded. You have to know life and business, having sat on all sides of all tables, from a warehouse to a boardroom. At the same time, lengthy experience does not guarantee becoming well-rounded. Some individuals may have many years of experience on their resumes, but they did not learn essential lessons or even picked up on the wrong ones.

It is usually noticeable when talent is encountered. When you listen to Bach's or Vivaldi's music, you can feel its spotless beauty. You experience the ultimate depth and integrity and perceive its eternal soul (Van Hoef, 2013).

You can watch a show of Timmy Trumpet at an electronic music festival and see how he is masterful and organic in putting up the show and entertaining massive crowds into a fandom (Tomorrowland, 2018).

You go to a Cirque du Soleil show and enjoy the inspirational performance staged by professional and talented athletes and entertainers (Cirque du Soleil, 2022).

Why can't we expect the same level of performance at other jobs, crafts, trades, and ventures? What is the difference? Why not make it an ambitious priority in life? Why not strive to become the best in the field of choice? Why not start asking these questions? Why not begin challenging your purpose in this regard?

Unexpectedly for many, AI comes to help address these questions and instigate the soul-seeking endeavor.

A big problem in IT is getting too specialized and narrowly focused on specific niches, often failing to grasp the entire scene where a performance happens. At the same time, those techies at the beginning of their careers who get to higher administrative levels prematurely quickly lose traction with the reality of how technologies work and better ways to organize them and miss out on this perspective.

It is a hard reality in IT that technologies are becoming increasingly complex and intertwined, and mastering them requires much time and dedication. A deep attachment to them and drowning and immersing oneself in them in a particular area of specialization is difficult to wrestle with.

Organizing an outstanding performance on a music stage is a unique combination of hard work and artful skills. It can get no less hard than in technologies to grasp and account for many details and the entire program that make the show a success. It is a complex endeavor.

There is another pitfall. Someone can have a broad outlook and hypothetically learn all the technologies in the world and even use them all but never reach a premier level. So many components contribute to that, and it is easy to miss some of them and derail.

Could it be that technology might not be suited for everyone who ventures into it casually? Not everyone who enters the world of writing and playing music achieves recognition and reward for their talent. Those who do not may choose to continue to pursue music as a hobby or as amateurs for enjoyment.

For example, I have a friend who aspired to become a prominent musician and ended up opening a restaurant where he plays with a band for patrons. He enjoys it a lot in this way, and patrons enjoy his joint as well!

Who am I referring to as a "celebrity techie," a particular "celebrity professional" category?

An engineer who is a talented artist who stages a beautiful show with his work (not just by its appearance but long-lasting results), who is proud of his artful work, while his audience is consistently delighted.

By show here, I do not mean making it a magician's performance with visual effects and communicative tricks, after which, when aromatic candle smoke settles, the client discovers a different kind of smoke from things burning down.

Where can a celebrity techie help? Here are some frequent situations where organizations experience pain points. However, this is not an exhaustive list. There can be many more challenging situations and more nuanced contexts.

Technology strategies, roadmaps, and architectures might be developed ad hoc or in the wrong direction, producing a technology monster inside the organization that is difficult to deal with and impacts everyone as an obligation rather than an enabler.

As noted earlier in this book, technology expertise is crucial for business executives and board members, as the accountability for technological errors and shortcomings ultimately rests with them. Technology investments and expenses are frequently substantial and keep increasing, accounting for up to 20% of revenue or more in some industries (Weins, 2020). These investments are expected to yield healthy profits while minimizing technical debt, affecting organizations' continuous performance.

Techies, including those at the helm as CIOs and CTOs, sometimes get overly excited about certain technologies without careful consideration from a business perspective, for their own self-development, due to missing knowledge, or by overlooking and not grasping critical business and technology considerations. As a result, an organization is left with bulky, cumbersome solutions that are difficult to support and use over time.

Vendors may shift their focus from the problems they are trying to solve to promoting their solutions for all use cases, regardless of their suitability. Independently assessing the suitability of third-party solutions and services and their alignment with the organization's technology and business strategy can be a complex undertaking with a significant impact on the business.

Vendors are inherently interested in selling their value-added proprietary solutions with a higher margin, even though it may be in their client's best interests to stick with a leaner technology stack and avoid huge excessive outlays, overcomplication, and lock-in.

Many clients start perceiving some cloud and SaaS models as excessively priced and looking for alternative options.

Coherent and agile teams of top-tier techies equipped with AI can now compete with vendors to create more economical, flexible, and better-suited custom solutions for businesses. Vendors would have to take notice.

Even in the case of AI, individuals and organizations have the option to run their own AI systems using open-source software libraries and models. By self-hosting and training these systems with their own data, they can avoid dependency on and data sharing with ChatGPT or similar SaaS AI services.

A viable strategy for vendors remains to increasingly embrace open-source practices while concentrating on client interaction and services. By doing so, they can enhance the usability of their products and make them more robust and competitive through exposure to public scrutiny.

It is worth mentioning that whatever technology options are considered for practical use cases, there is rarely a clear-cut winner. Due to a combination of reasons, most technologies we have for practical use are far from ideal, and any option being considered has its own compelling advantages as well as disadvantages. **The choice is often an artful, hard, and carefully managed compromise.**

Arguments such as "Which is better in martial arts—karate or kickboxing?" are more intense and extend to a wide range of topics and options in the technology arena.

A significant challenge within organizational dynamics is the response of many technology practitioners to the profession's "stress test." Overwhelmed by the high demand for technical expertise, the need for continuous learning, and the pressure to stretch beyond their capabilities to deliver results, they often conclude that ascending to a leadership tier is the solution to mitigate these pressures. And since it can be difficult for many organizations to assess the level of techies, at least over the initial (sometimes quite lengthy) period of employment, they manage to move on this plan up the ladder.

The issue with this is that, on one side, it dramatically diminishes the productive technology workforce who can do hands-on work, who "do the stuff": designing, developing, testing, and implementing. On the other side, this creates bloated and increasingly detached administrative levels of managers without sufficient expertise because they skipped earning that expertise early in their careers. These two centrifugal forces are significantly eroding the profession, productivity, and output in IT.

This becomes a shaky construct. Many technology managers are sticking to primarily administrative and bureaucratic duties, leaving most of technology governance liberty to the doers who look at things from their own often disjointed perspectives, making suboptimal and even counterproductive decisions impacting the overall technology stance of an organization.

The lack of subject matter expertise at managerial levels negatively impacts the quality of outcomes, hiring practices, and overall performance of organizations.

Technology managers who largely skipped accumulating technology expertise become commoditized and vulnerable to layoffs. So, this situation gets really dramatic from all angles.

Another issue arises when different technology teams, speaking in "different languages," both technology-wise and business-wise, fail to integrate their efforts and work products effectively.

Burnout problems in the IT professions can actually be solved with premier-level techies. With their highly intensive and outstanding work associated with corresponding compensation levels, they should be able to choose to retire earlier. Mastery and great motivation can also significantly mitigate and counterbalance burnout.

Digital products can now be created at totally different quality levels, and organizations can finally get the opportunity to innovate more easily without a heavy anchor of technical debt restraining them.

At the same time, under-qualified practitioners weaponized with AI will be able to exacerbate the technical debt to new levels. This is one of the greatest risks of using AI.

Avoiding such issues and helping an organization to charter the right technology course requires a premier level of qualifications.

By helping organizations turn technology around from an anchor and necessary evil into gold, a celebrity technologist enables corporate leaders and organizations to save significant money and increase earnings in their respective sectors.

As such, a celebrity-level professional can achieve new heights in compensation.

In the old paradigm, you were expected to do your job. AI enables premier-level professionals to do an artful job, not just a good job. It raises the bar.

The making of a celebrity techie

The training of a celebrity professional is akin to training to become a champion in a professional sport or to become a Jedi, figuratively speaking. As mentioned earlier, it takes a lot of concentration and effort.

Modern academia alone cannot provide that. Its long-standing issue has been a lack of lecturers with hardcore practical experience. Yes, some lecturers get on company boards. Also, from time to time, universities invite visiting lecturers from the industry. However, that only changes a little about the overall situation.

So, first, traditional education often lacks a comprehensive, up-to-date, and practical grasp of technologies.

Second, and this is a very important factor, there are not so many people in both business and academia who have a talent for teaching and can expedite professional training and comprehension with clear, structured, and systemic explanations. You can rarely see courses in academia that are taught this way, and they become gems of their hosting institutions.

The technology industry changes so much and so fast that courses need to be updated as much and as fast on a rolling basis. Not just courses themselves but their combinations into programs. Programs need to be continuously revised.

It gets common for new graduates to face difficulties in securing a job in their field of education or beyond. For example, young Australians are voicing concerns about finding a job after graduation (Madigan, 2024). New York Fed reported in 2017 that about 34% of college graduates in the US were underemployed, working in jobs that do not require a college degree (Cooper, 2017). Even considering that this statistic includes degrees in subjects that are not in demand, it casts a shadow on the overall function and value of college education.

It is not that colleges and universities cannot be relevant in the new realities. They would rather need to become leaner, more agile, and more vigorous in offering relevant education.

What is positive about traditional formal higher education is that it teaches discipline in pursuing your goals, in study, and then in your career.

What would the training program need to be for a future top professional? The critical components of the training would need to include the following:

1. Make up your mind about your choice of profession and start on that as early as possible in your life.

2. Learn and practice non-stop.

3. Get premier league coaching in your profession.

4. Work on your achievements and their contribution to your brand. The coaching that you take will be a part of your brand.

5. Stay tuned for changes and keep advancing.

It works pretty much the same as in professional sports.

I have honed the methods of delivering hyper-intensive advisory, training, and coaching on technologies in the current dynamic AI-driven environment, and they are thoroughly tested and proven to be highly effective. At CelebrityTechie.com, we are shaping an Academy of the future to nurture star professionals while also providing coaching, training, and consulting services to top business leaders and executives.

How to recognize a celebrity techie

A big organizational challenge when working with techies is difficulties in assessing their level and suitability for roles. How can you recognize that you are dealing with a premier-level professional if you are a technologist specializing in different areas or a non-techie?

The secret lies in testing the candidates' limits and understanding their authentic potential in the complex. For this, you have got to be prepared to go broad and in-depth on a virtually unlimited scale.

Another important factor is hands-on expertise, which is crucial across all occupations and levels. It is not uncommon for candidates applying for senior positions to lack sufficient real-life business and technology knowledge and experience, often focusing solely on the intention to "manage."

This works out poorly, and if such individuals get into senior managerial positions, they continuously experience stress and discomfort. They are also easily confused and manipulated by "interest groups" among reports and other stakeholders who are more immersed in the situations "on the ground."

Involving trusted premier-level techies in doing candidate assessments, especially for critical roles in an organization, allows to minimize losses where a premium can be paid to someone who might not be up to that level, and time can be lost, along with other costs borne.

A single comprehensive interview can often yield definitive conclusions. Even candidates who are not selected frequently find that such interview experience provides fresh perspectives and motivates them.

How to work with celebrity techie

Accountability and flexibility are the critical components engrained into a celebrity techie's personality and nature of work.

Reputation is a big factor for a celebrity techie and immensely contributes to accountability. A premier league professional cannot drop the ball here and there. He or she is expected to consistently perform at the top level. Dents of failing to deliver on commitments and expectations can cause a celebrity techie to lose his champion title.

From this perspective, a celebrity techie is highly self-motivated. Constructive communications and clearly formulated work goals help establish accountability for the engagement.

The material component is a key motivator and an essential factor, among other considerations. Executive-level compensation and participation in ownership are fair prizes for proven results and contributions.

A premier league professional works super hard. With all the precautions and following work safety rules, it can affect health. For example, vision can be affected, as prolonged working with screens is unavoidable for techies. Long hours can affect work-life balance, and compensation can reasonably mitigate that and other adverse effects over the long term, embracing the possibility of fully financed early retirement if needed.

Flexibility and hybrid work modes are also crucial in the premier league. Lack of accountability is what usually taints the hybrid work modes, but that should not be a problem for a premier-level professional.

It is crucial to maintain active communications with parties involved in an undertaking, and it is equally important to be able to step back and focus on analyzing and processing the information as needed. Cal Newport's book, "Deep Work," highlights the importance of uninterrupted concentration and its positive impact on producing high-end results. A typical office environment often does not provide sufficient opportunities for complete concentration without distractions (Newport, 2016).

Working with a premier-level professional should be a joyful collaborative journey of problem-solving and creativity.

The making of a celebrity professional is a laborious and comprehensive process that requires a lot of dedication and focused training at a level similar to that in professional sports.

References

Van Hoef, G. (2013, June 9). J.S. Bach - Toccata & Fugue in D-minor BWV 565 - Stephanuskerk Hasselt [Video]. YouTube.

Tomorrowland. (2018, July 24). Timmy Trumpet | Tomorrowland Belgium 2018 [Video]. YouTube.

Cirque du Soleil. (2022, January 28). Alegría | Show Trailer | Cirque du Soleil [Video]. YouTube.

Weins, K. (2020, March 5). IT spending by industry. Flexera.

Madigan, M. (2024, March 6). Young Aussie reveals just how hard it is to get a job even with a university degree. news.com.au.

Cooper, P. (2017, July 13). New York Fed highlights underemployment among college graduates. Forbes.

Newport, C. (2016). Deep work: Rules for focused success in a distracted world. Grand Central Publishing.

HOW EVERYONE IS AFFECTED AND HOW TO PREPARE

It is global, it is pervasive, and everyone is affected.

The ever-growing use of AI will make the current economic model unsustainable and plainly unworkable. Many processes can already be facilitated with AI and automated to a high degree.

Andrew Yang quoted in his book an operations manager from a major tech company who said: "I just helped set up a factory that had 70 percent fewer workers than one even a few years ago would have had, and most of them are high-end technicians on laptops." (Yang, 2019)

An article on the Time website admits that automation has been around for centuries, but it also points out some growing worries about what is happening these days (Semuels, 2020). They specifically mention that in the past, the U.S. dealt with technology changes by investing in education, and workers had more time to retrain and get used to the changes that automation brought. Also, companies used to create new jobs to replace the old ones almost evenly. But now, changes are happening so fast that both retraining and creating new opportunities are increasingly falling behind. Plus, the article points out that tax policies are encouraging even more investment in automation.

This article is among the countless publications on this subject that continue to grow fast in number.

Our society has to adapt and change its paradigms to accommodate the diminishing need for jobs in their current forms in offices and elsewhere. Reorienting and retraining in the new realities become vital.

Of course, being a professional in knowledge-based occupations is not the only option in the new economy. Alternative career paths include entrepreneurship and skilled trades, both of which increasingly leverage AI while highlighting the crucial role of humans. Entrepreneurship enables you to create opportunities for yourself and others, while skilled trades have ongoing shortages (Meacham, 2022), offering the potential to earn well and even combine this work with entrepreneurship.

Some people would want to try various kinds of hustling, including trading and investing, e-commerce, copywriting, entertainment, etc.

Whatever you do, the same approach applies—you must try your best and be persistent while being flexible and able to manage risks.

Employing humans in some occupations may soon be considered a differentiator. For instance, in customer support, it is often more comfortable to realize that you talk to real humans who can empathize with a situation and are meaningful to talk to. Also, do not forget about AI hallucination phenomena; they are real and bitter when encountered. Even with the rapid advancements in AI technology, many people prefer to speak to a real person rather than an algorithm with no real emotions.

Certain professions where a lot is at stake, such as judges in courts, should mandate the employment of people as primary decision-makers. This is because human judges are able to take into account a range of complex factors beyond what can be easily programmed into an algorithm or machine learning model. In addition, human judges are also able to apply interpretative reasoning to legal disputes and make use of their experience and expertise to ensure proper judgment and accountability.

Just recently, the European Parliament approved the Artificial Intelligence Act that stipulates obligations for high-risk systems, among others:

"Clear obligations are also foreseen for other high-risk AI systems (due to their significant potential harm to health, safety, fundamental rights, environment, democracy and the rule of law). Examples of high-risk AI uses include critical infrastructure, education and vocational training, employment, essential private and public services (e.g., healthcare, banking), certain systems in law enforcement, migration and border management, justice and democratic processes (e.g. influencing elections). **Such systems must assess and reduce risks,**

maintain use logs, be transparent and accurate, and ensure human oversight. Citizens will have a right to submit complaints about AI systems and receive explanations about decisions based on high-risk AI systems that affect their rights. (European Parliament, 2024)"

Overall, the retention of certain occupations for humans is a potential solution to ensure that the benefits of automation are balanced with the preservation of human jobs and the need for human decision-making in certain settings.

Other options, such as a shorter work week, time banking (essentially labor bartering), engaging in rural lifestyles, and UBI, are a significant departure from the current economy and society models and do not seem to be taking off on a substantial scale.

So, what to do?

Well, you need to think seriously about your options and what opportunities you can target with your capabilities and resources. A number of viable options are considered throughout this book.

Once you have a clear idea of your objective, it is time to act. This can involve acquiring new skills, networking with others in your field, or experimenting with different approaches to get a sense of what works best.

By taking proactive steps to prepare yourself, you can increase your chances of success and make the most of your efforts.

Are you still committed to becoming a star professional in your field? Do you believe you have the capability to aim for the highest echelons in the profession? What do you think is missing that you need to achieve that goal?

Consider the expanding role of technology expertise in today's workforce across all occupations, particularly at leadership levels. Historically, the primary criteria for leadership positions were people skills. While these skills remain crucial, they no longer guarantee long-term job security. Technology expertise is rapidly becoming essential in executive management and on boards of directors. It is not merely a differentiator for executive leadership but a vital necessity that prevents leaders from becoming commoditized and easily replaceable.

AI by itself will not make you a star professional. It can actually play out worse for you. AI does not spare you from learning your field at a deep level and engaging with original content.

Remember, the European Parliament considers education a "high-risk AI use," and for a good reason.

First, employing AI actively in education without a deep understanding of the subject may foster a superficial approach to a subject area and hinder the development of essential learning and analytical skills. As Henebery (2023) points out, "The fact that the answers provided by LLMs are so good (at least superficially) leaves no room for the student to interpret and understand."

Second, AI is easily influenced by the information used for its training, which can result in a distorted and patched understanding of the subject area. Without directly reviewing sources, a student cannot be confident about the quality of knowledge obtained from AI.

Artificial Intelligence actually raises the bar for you, requiring you to have a more-dimensional, higher knowledge ("uber knowledge") in the profession to be able to steer the AI in the right direction and, at the most basic level, distinguish between good and bad outcomes from AI.

It is as if you have become the captain of a modern cruise ship or the pilot of a supersonic jet, now responsible for steering it. Even with numerous support systems and automation, you are still required to have extensive knowledge to operate them effectively.

Alongside the increasing demand for more profound knowledge, professionals must also learn and master relevant technologies to excel in their respective fields.

The world is still largely oblivious to the rapid proliferation of AI, its implications, and its consequences. To a significant extent, this is due to the "Sirens" of various industries, whose voices lull us into a false sense of security, for example (Steinhardt, 2023). In Greek mythology, the Sirens famously enticed sailors with their captivating music and sweet voices, only to lead them to disaster on the rocky shores of their island. Similarly, today's Sirens assure us that AI will not lead to job losses but will instead automate routine tasks and allow people to focus on the creative aspects of their work.

However, this is only a fraction of the real story, as this book demonstrates. AI has already caused and will continue to cause significant displacement across industries (Constantz, 2024). The transformation is rapid and not as smooth as it could be desired, with a pressing need for reskilling and adaptation in response to the increasing integration of Artificial Intelligence and robotics in the economy. For example, it is estimated that a staggering minimum number of 1.5 million workers across all industries in just a single Australian state will need to be reskilled in just about five years (Enciso, 2024).

This book aims to draw attention to these topics, which are important for everyone, so that everyone can start thinking about how they will handle these changes and begin planning for them.

The actions we take as individuals in the era of AI will significantly shape our own future and that of society. To navigate this new reality, it is crucial to continuously learn, remain adaptable, and strive for excellence in what we do. It is essential to stay informed about emerging technologies, their best use and implications, and we must be proactive in acquiring new skills and knowledge in areas where AI complements human capabilities. By doing so, we can leverage AI to enhance our lives and create a future that reflects our collective aspirations and values.

References

Yang, A. (2019). The war on normal people: The truth about America's disappearing jobs and why universal basic income is our future. Hachette Books.

Semuels, A. (2020, August 6). Millions of Americans have lost jobs in the pandemic—And robots and AI are replacing them faster than ever. Time.

Meacham, S. (2022, August 10). Massive tradie shortage puts more pressure on Aussie construction industry. 9News.

European Parliament. (2024, March 13). Artificial Intelligence Act: MEPs adopt landmark law. European Parliament News.

Henebery, B. (2023, May 2). Generative AI: The risks, strategies, and opportunities for schools. The Educator Magazine.

Steinhardt, S. J. (2023, August 9). Big Four agree: AI will not replace accountants. The Trusted Professional: The Newspaper of the New York State Society of CPAs.

Constantz, J. (2024, February 9). AI is driving more layoffs than companies want to admit. Bloomberg.

Enciso, C. (2024, April 3). Over 1.5 million WA workers will be substantially impacted by critical technology: ACS report. The West Australian.

Made in the USA
Monee, IL
04 November 2026

67391559R00046